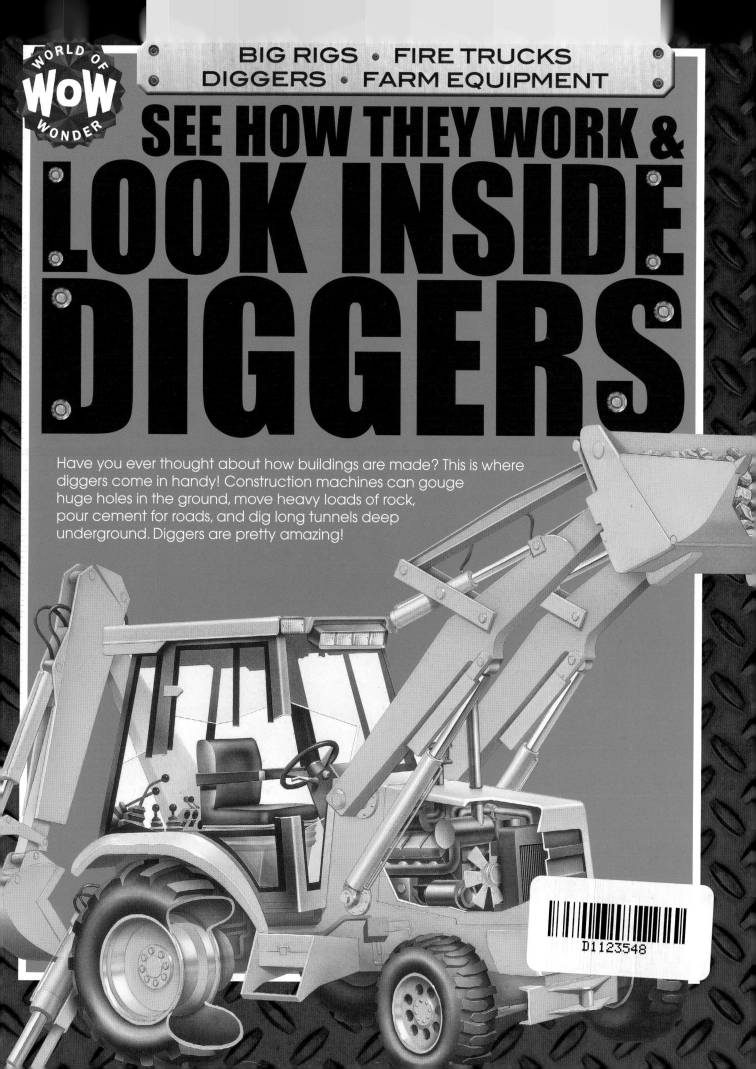

WORLD OF WOW WONDER

BIG RIGS • FIRE TRUCKS
DIGGERS • FARM EQUIPMENT

SEE HOW THEY WORK &
LOOK INSIDE
DIGGERS

Have you ever thought about how buildings are made? This is where
diggers come in handy! Construction machines can gouge
huge holes in the ground, move heavy loads of rock,
pour cement for roads, and dig long tunnels deep
underground. Diggers are pretty amazing!

© 2015 Flowerpot Press

Contents under license from Aladdin Books Ltd.

Flowerpot Press
142 2nd Avenue North
Franklin, TN 37064

Flowerpot Press is a Division of Kamalu LLC, Franklin, TN, U.S.A.
and Flowerpot Children's Press Inc., Oakville, ON, Canada.

ISBN: 978-1-4867-0564-1

Editor: Michael Flaherty

Design: David West Children's Book Design

Designer: Simon Morse

Illustrators: Simon Tegg & Ross Watton

American Edition Editors: Johannah Gilman Paiva
                                        and Ashley Rideout

American Redesign: Stephanie Meyers

Consultant: Robert Lawrence Hendrick III, EIT

Printed in China.

TABLE OF CONTENTS

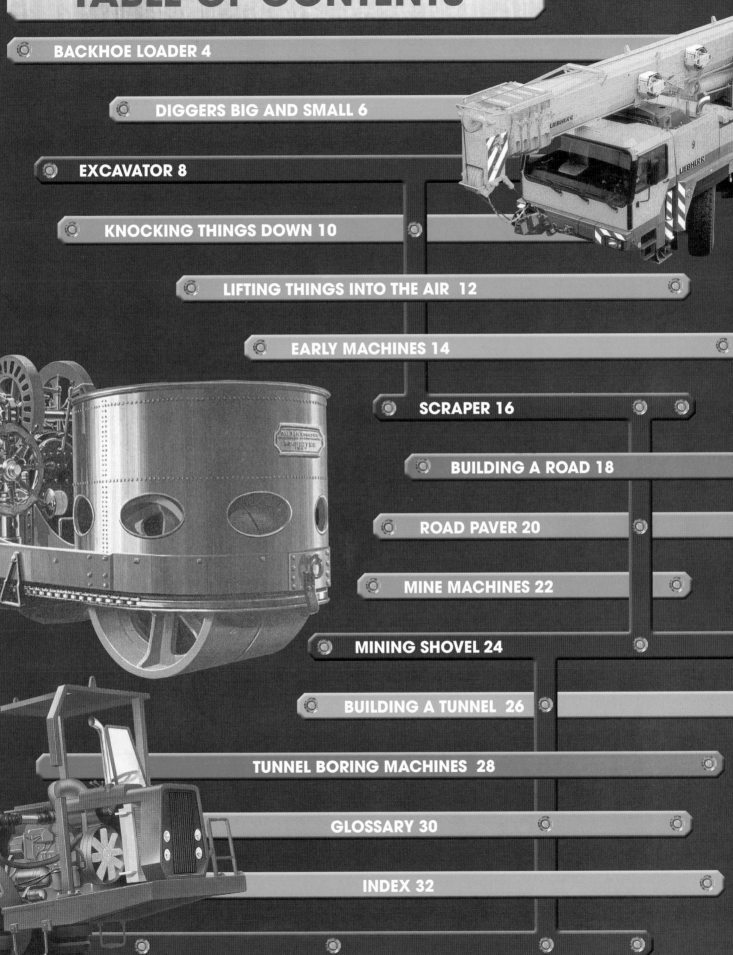

BACKHOE LOADER

This unusual-looking machine is possibly the most useful digger on any construction site: the backhoe loader. It has tools fitted to its front and back. On the front is a huge scoop, or "bucket," that can be used to scoop up dirt, rocks, or other rubble. On the back is a movable arm called a "backhoe." This can be used to dig trenches or pull down buildings. The backhoe loader can also be fitted with different tools, such as a pneumatic drill (see page 18) and a mechanical claw, in order to do other jobs that are needed around the work site.

Front bucket
The front of the bucket is hinged so the rubble can be easily dropped into a waiting truck. The bucket on this backhoe loader can lift nearly 3 tons (2.7 metric tons) of rock! That's about as much as three orca whales!

Cab
The cab is fitted with large glass screens to give the driver a good view of the backhoe's arm.

Stabilizers
These special arms reach out to take the weight off the back wheels and provide a steady platform when the backhoe is being used.

Chunky tires
The backhoe loader has large, chunky tires so it can drive over rough ground.

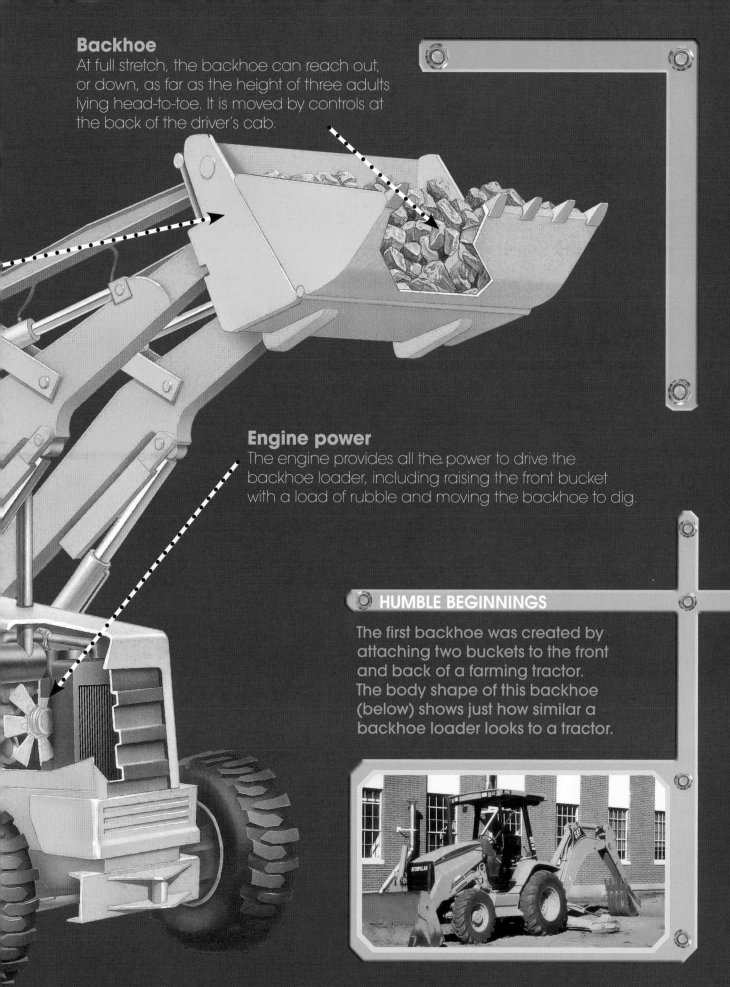

Backhoe

At full stretch, the backhoe can reach out, or down, as far as the height of three adults lying head-to-toe. It is moved by controls at the back of the driver's cab.

Engine power

The engine provides all the power to drive the backhoe loader, including raising the front bucket with a load of rubble and moving the backhoe to dig.

HUMBLE BEGINNINGS

The first backhoe was created by attaching two buckets to the front and back of a farming tractor. The body shape of this backhoe (below) shows just how similar a backhoe loader looks to a tractor.

DIGGERS BIG AND SMALL

Diggers come in all shapes and sizes. Some need to be tall and strong to lift massive loads, while others need to be small and nimble for carrying small loads around tight corners and narrow passages. Some move on wheels, while others use caterpillar tracks to move themselves around. No matter the job, there's a digger out there made for it!

DREDGER

Not all diggers are used on land. Sometimes waterways get blocked by too much sand and dirt building up.
If the waterway gets too shallow, boats and ships may not be able to pass through it. That's where the dredger comes in handy.
A dredger (right) digs up mud from the bottom of the sea, or a river, and puts it on a boat to be hauled away. This prevents the waterway from getting blocked.

FOUNDATION PILES

A foundation is the base of a building. When a tall building is built, it needs a deep foundation so it won't sink into the ground. The taller the building, the stronger its foundation needs to be to support all the weight. Special drill rigs fitted with huge drill bits (above left) dig holes into the ground in which the foundation piles, or supports, are placed to make a big, strong foundation.

SKID STEER

This small building machine (right) is called a "skid steer." It is very useful in small spaces because it can move things without needing much room at all to turn around. It changes direction by speeding up the wheels on one side faster than those on the other side, spinning the machine around.

MINI EXCAVATOR

A mini excavator (above right) is used to dig where space is tight. With its small arm and bucket, it can dig very narrow trenches. The body of the excavator above the tracks can rotate in a full circle. Its small size and light weight make it easy to transport and perfect for jobs in small spaces.

EXCAVATOR

These mighty machines are seen on the largest construction sites. They are often used to dig the foundations for buildings or to dig wide trenches. Their strong hydraulic arms are also used to pull down buildings, or they can be fitted with enormous claws to pick up heavy objects. Beneath the driver's cab are two large caterpillar tracks. These spread the weight of the excavator over a larger area to stop it from sinking into soft ground. The largest excavators can weigh 80 tons (72.6 metric tons)—that's as much as 25 fully grown elephants!

Cab

Engine
The engines used to power excavators need to be very strong—about four times more powerful than the engine used to run a car!

Caterpillar tracks
Caterpillar tracks have ridges to stop vehicles from slipping on soft or icy ground.

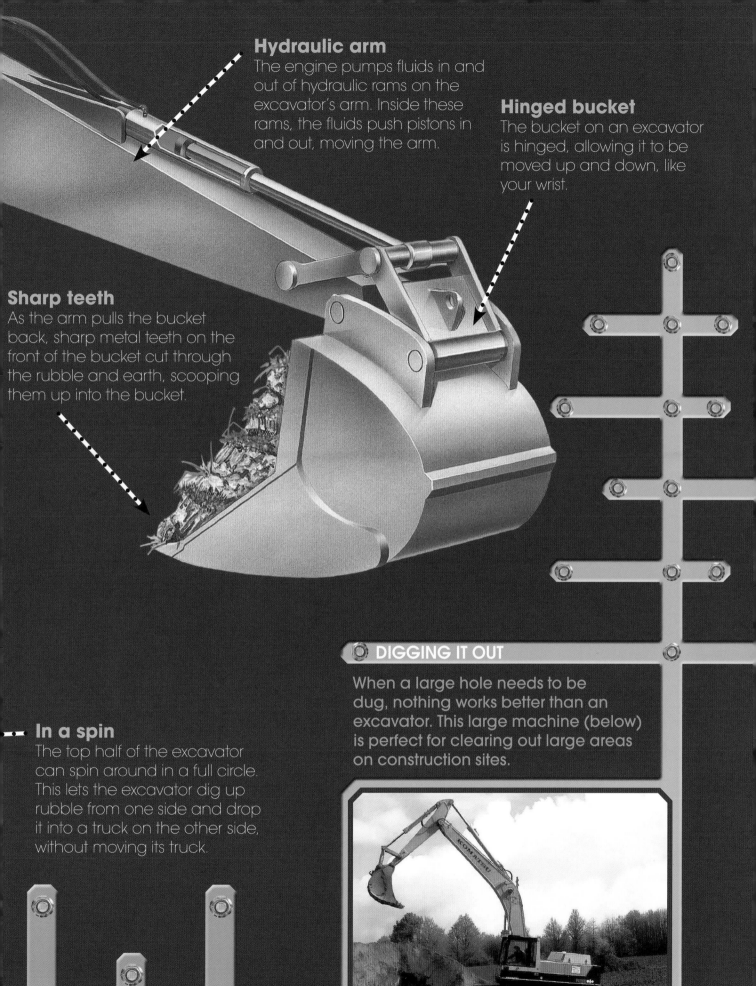

Hydraulic arm
The engine pumps fluids in and out of hydraulic rams on the excavator's arm. Inside these rams, the fluids push pistons in and out, moving the arm.

Hinged bucket
The bucket on an excavator is hinged, allowing it to be moved up and down, like your wrist.

Sharp teeth
As the arm pulls the bucket back, sharp metal teeth on the front of the bucket cut through the rubble and earth, scooping them up into the bucket.

In a spin
The top half of the excavator can spin around in a full circle. This lets the excavator dig up rubble from one side and drop it into a truck on the other side, without moving its truck.

DIGGING IT OUT
When a large hole needs to be dug, nothing works better than an excavator. This large machine (below) is perfect for clearing out large areas on construction sites.

KNOCKING THINGS DOWN

Not all diggers are used for creating buildings. Sometimes a building gets so old that it is no longer safe to use. When this happens, special demolition machines are used to tear the building down so that something else can be built. Large buildings can be taken down within minutes without disturbing a single building beside them!

GOING, GOING, GONE

Not all buildings can be demolished cheaply and quickly by machines. Very tall buildings may need dynamite to knock them down. Once the building is completely empty, sticks of dynamite are put in the places in the building where they will have most effect (top left). When they explode, they cause the building to collapse (bottom left). It takes skill to demolish a building without damaging surrounding buildings or spreading rubble everywhere(bottom right)!

BULLDOZER

When a building is knocked down, the rubble has to be cleared away. Bulldozers (left) are fitted with huge blades at the front to push the debris to one side of the site where it can be loaded into trucks.

GRAPPLE TRUCK

The claw on the end of this machine (left) is used to pull down buildings. The driver's cab is fitted with strong bullet-proof glass to protect the operator from falling rubble.

LIFTING THINGS INTO THE AIR

Have you ever wondered how skyscrapers are built so tall? To get the materials up to the top, workers use a crane to lift them high in the air. The crane has a hook at the end that attaches to beams and other important building materials and hauls them to the top of the building. The first cranes were used as early as the 6th century in Ancient Greece. In 1480, the Italian inventor and artist Leonardo da Vinci designed the first rotating crane. Today, the tallest mobile crane in the world is the Rosenkranz K1000. It weighs nearly 900 tons (816.5 metric tons), but it can lift nearly 1,120 tons (1,016 metric tons). Its arm can reach a height of 663 feet (202 meters).

TOWER CRANE

The tallest cranes, tower cranes, are put together at the work site in stages (left). They have working arms, or jibs, that are attached to the mast, or tower. Loads are lifted by cables and a hook from one end of the jib, while heavy weights on the other end stop the crane from falling over.

EYE IN THE SKY

The crane driver sits in a small cab at the top of the crane (right). From here, he or she controls where the crane's arm is pointing and how much cable is needed to lift or lower the load.

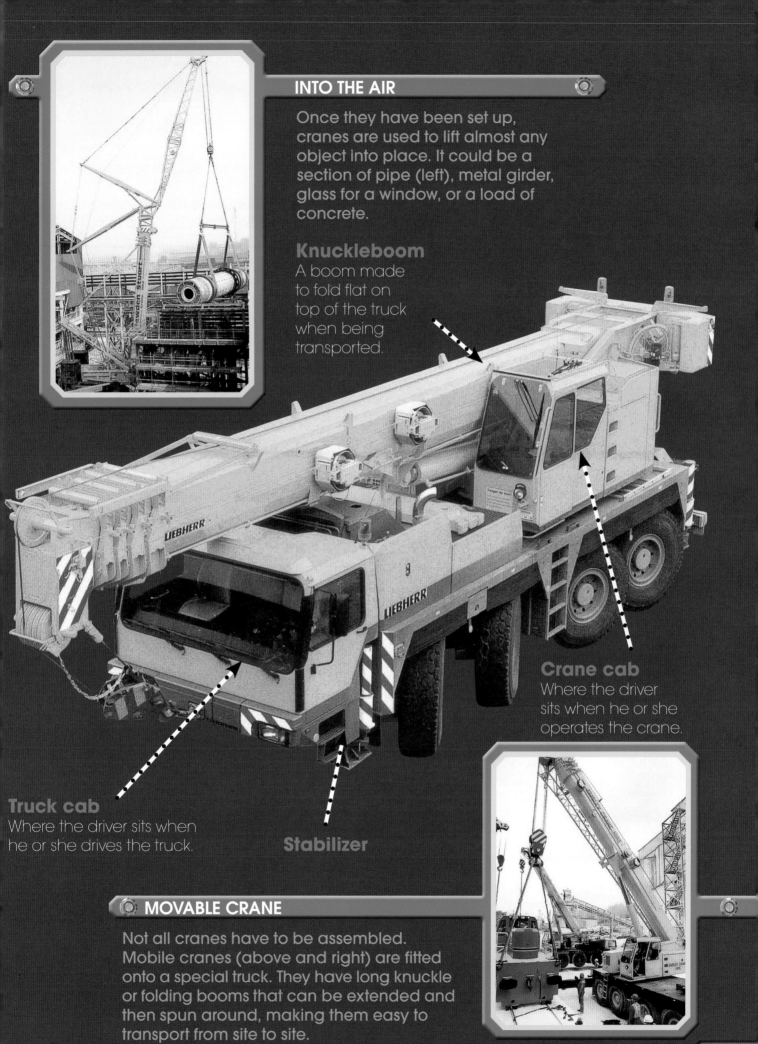

INTO THE AIR

Once they have been set up, cranes are used to lift almost any object into place. It could be a section of pipe (left), metal girder, glass for a window, or a load of concrete.

Knuckleboom
A boom made to fold flat on top of the truck when being transported.

Crane cab
Where the driver sits when he or she operates the crane.

Truck cab
Where the driver sits when he or she drives the truck.

Stabilizer

MOVABLE CRANE

Not all cranes have to be assembled. Mobile cranes (above and right) are fitted onto a special truck. They have long knuckle or folding booms that can be extended and then spun around, making them easy to transport from site to site.

EARLY MACHINES

How people built things changed dramatically with the invention of machines. Many early machines were powered by steam. Even though building is hard work, it is much easier with the help of machines!

PICK AND SHOVEL

Today, a single machine operated by one person can do the same work that used to take many people equipped with picks and shovels (right).

FULL STEAM AHEAD

Before powerful internal combustion engines and electric motors were invented, building machines were powered by other means. This shovel (below) was powered by a steam engine.

STEAM ROLLER

Like the steam shovel (above), this steam roller (right) was powered by a steam engine. The roller's huge, heavy wheels were pushed forward by energy created in the boiler. The steam roller was used to level surfaces. From roads to airfields, this machine made the ground level so that other vehicles could have a smoother ride. Created in France in 1860, the steam roller worked by using its heavy weight to smash down the ground (often asphalt), while the wide, round rollers made the surface smooth. Machines like the steam roller paved the way for new ways of building, bigger and faster

⊙ TRENCH DIGGING

This machine, called an "endless bucket trencher," used a series of small buckets on a belt to move earth. As this belt moved around, the buckets scoop out the earth to dig a trench. Bucket trenchers, like this one from 1901 (right), were used for many years, long before the invention of hydraulic arms (see page 9).

Funnel
Allows extra smoke to leave the boiler area.

Boiler
Where water is heated to power the steam roller.

Wheel
These large wheels were designed to flatten asphalt to prepare a road for cars to drive on it.

Steering wheel
This machine was directed by a large wheel like those on a ship.

SCRAPER

This machine is called a "scraper" because it is used to scrape up layers of earth, preparing the ground for construction, road making, or farming. It has a huge blade underneath that digs up the soil as the scraper drives forward. The soil is then collected in a huge container and carried to another site where the scraper can drop it. Sometimes, even the two engines in a scraper are not powerful enough and it may need to be pushed by another vehicle.

Rear engine
This scraper is fitted with two engines. The engine at the rear drives the rear wheels, pushing the scraper along.

Chunky tires

Collecting the load
The container on this scraper can hold about 17 tons (15.4 metric tons) of earth—nearly the weight of 250 adults!

Blade
The scraper's blade can be pushed up to 12 inches (30 centimeters) into the ground to collect large amounts of soil to be moved.

Hinged front
The tractor is linked to the rest of the scraper by a large hinge that bends to make turning easier.

Tractor
The front of the scraper, which contains one engine and the driver's cab, is called the "tractor."

Powerful engines
In order to push and pull the scraper's blade through the ground, the engines need to be very powerful. Each engine of this scraper is as powerful as the engines from ten cars! Having two separate engines allows the scraper to power the two separate sets of wheels to provide better traction.

PREPARING FOR THE ROAD
Scrapers are heavy, hardworking, efficient machines that work on rough terrain to make it ready for the work that needs to be done, like preparing land for a new road!

BUILDING A ROAD

There are many important stages to building a road. First, the ground has to be prepared. A grader will compact the soil to flatten it down. Then, gravel will be placed as a base for the road. Next comes the asphalt paver, which drops the asphalt onto the ground. A roller comes behind the paver to flatten the asphalt and make sure it will be a nice, smooth road. Once the asphalt has dried, workers will paint lines on the road with a special truck.

DIGGING IT UP

Sometimes old roads need to be dug up before you can make a new one. This backhoe loader has been fitted with a pneumatic drill (left). In the pneumatic drill, air is squashed very hard, which forces a piston up and down. This piston smashes onto the drill, hammering it into the road, and breaking up the old asphalt.

FINE GRADES

A grader is fitted with an angled blade. As the grader drives along, this blade pushes the soil to one side to create the flat, even surface. This is called "grading." It can also be used to create even slopes (below).

SQUISHING IT FLAT

After the asphalt has been laid (see pages 20-21), a roller (right) is used to squash it flat and make it smooth. The heaviest roller can weigh 35 tons (31.8 metric tons)—that's the same as 22 adult giraffes!

EXCAVATOR

As well as demolishing buildings and digging trenches, excavators can be used to build roads (left). Their buckets can be used to move rubble and to grade soil to create an even surface.

ROAD PAVER

Once a pathway has been dug for a road, the road's foundations are built by putting down layers of materials. The final layer is built by an asphalt paver. This noisy, smelly machine moves forward slowly, spreading a thin layer of asphalt—a black, sticky substance mixed with stones. Asphalt can be squashed and shaped when it is hot and is rolled flat by a roller (see page 19) to make it smooth. When it cools, it sets hard enough to withstand heavy traffic. A paver is also used to replace old or damaged asphalt.

Exhaust
As the engine burns fuel to work, it creates waste gases. These gases are carried out of the engine along a metal pipe called the "exhaust pipe."

Engine

Hopper
Dump trucks empty the hot asphalt into this large hopper at the front of the road paver.

Conveyor belt
This conveyor belt carries the asphalt from the hopper to the rear of the road paver.

ROAD PAVERS AT WORK

Asphalt is being added from a dump truck to the paver's hopper. Pavers have to work at a steady, consistent pace to be able to make the road as smooth as possible.

Controls

On some road pavers, the steering wheel and other controls can be moved from one side of the paver to the other. This lets the driver keep a close eye on either side of the road.

Giant corkscrew

At the back of the road paver is a large screw called an "auger." This auger spins slowly, gathering the asphalt that is being dumped onto it and spreading the sticky substance evenly across the road.

Sole plates

Once the asphalt has been spread over the road, special heated plates, called "sole plates," flatten and smooth the asphalt. Sometimes these sole plates can be extended so the paver can lay a wider road.

MINE MACHINES

Valuable minerals and metals come from deep in the earth. In order to gather these materials, people dig deep caves in the ground called "mines." The materials to be mined can be hard to get out. That is where these strong machines come into play! Mining machines can do the work that used to take days to do by hand. They make mining more efficient and less hazardous for the workers.

Cutting boom

This boom can be adjusted up and down to get the buckets at just the right angle so that they are low enough to pick up the material without getting stuck.

Counterweight

This excavator needs a heavy weight to prevent it from tipping over when the already heavy boom at the other end picks up a large load.

LIFE'S A DRAG

This massive digging machine (left) is called a "drag-line excavator." It throws out an enormous bucket on the end of a cable and then drags it back over the surface. As it is dragged back, sharp metal teeth on the front of the bucket scrape up the earth.

MASSIVE BUCKETS

The buckets on a mining excavator need to be big (right). The largest buckets are wide enough to hold two minivans!

Bucket wheel
This wheel has many buckets attached to it. As the wheel rotates, each bucket picks up a load and carries it to a conveyor belt located on the boom.

WHEEL AND BUCKET

This enormous machine (left) is called a "bucket-wheel excavator." Not all mines are hidden deep beneath the ground. Sometimes, minerals lie close to or on the surface. To get at these minerals, huge machines dig up the rock, creating enormous mines called "opencast mines." The excavator is fitted with a massive wheel ringed with buckets. The huge wheel at the front of the excavator spins around, and the buckets dig into the earth, picking up large amounts of material as they go by. One of these machines, called "Big Muskie," holds the world record for being the largest land vehicle, weighing 12,000 tons (10,866.2 metric tons)!

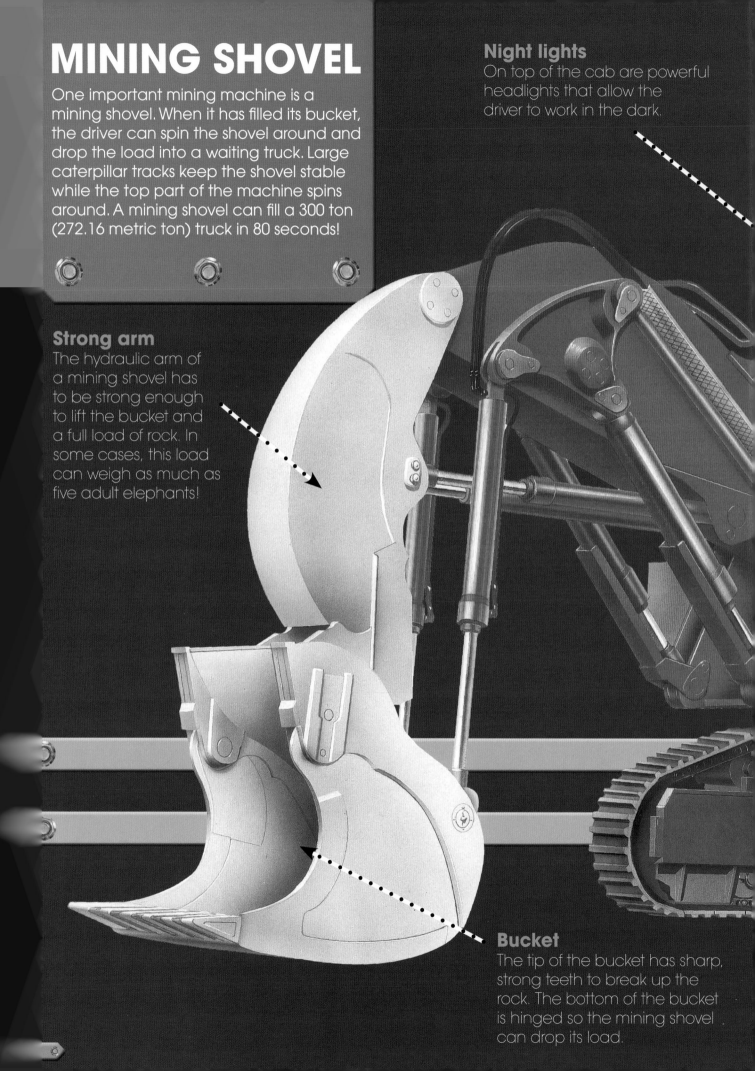

MINING SHOVEL

One important mining machine is a mining shovel. When it has filled its bucket, the driver can spin the shovel around and drop the load into a waiting truck. Large caterpillar tracks keep the shovel stable while the top part of the machine spins around. A mining shovel can fill a 300 ton (272.16 metric ton) truck in 80 seconds!

Night lights
On top of the cab are powerful headlights that allow the driver to work in the dark.

Strong arm
The hydraulic arm of a mining shovel has to be strong enough to lift the bucket and a full load of rock. In some cases, this load can weigh as much as five adult elephants!

Bucket
The tip of the bucket has sharp, strong teeth to break up the rock. The bottom of the bucket is hinged so the mining shovel can drop its load.

Swing gear

Underneath the driver's cab is the swing gear, which spins the mining shovel around in a complete circle.

Twin engines

Inside many mining shovels are two powerful engines, creating as much power as 35 cars. The shovel can still operate even if one of the engines breaks down.

Cab

BUILDING A TUNNEL

Sometimes traveling under the ground can be faster or more efficient than on open land. Large underground tunnels can be so deep that they go under the sea! Some special machines are used to build tunnels underground. One famous tunnel, the Mount Cenis Tunnel, runs under the Alps and was the first tunnel built with a pneumatic drill, which was invented in 1861 by the French engineer Germain Sommeiller. Another famous tunnel is the Channel Tunnel, a tunnel under the English Channel, which connects England to France. The tunnel-boring machines used to dig the Channel Tunnel were 49 feet (15 meters) long and weighed 1,300 tons (1,179.3 metric tons). Behind them were trains, each about 590 feet (180 meters) long and with over 1,000 tons (907.2 metric tons) of equipment.

LONG TUNNEL

The world's longest rail tunnel is the Seikan rail tunnel in Japan. It is 33.5 miles (54 kilometers) long. It runs under the sea between the main island of Honshu and the island of Hokkaido.

UNDER THE SEA

It took men and women using tunnel-boring machines over four years to dig the Channel Tunnel (see pages 28-29). This picture (left) shows the exciting moment when the two halves of the tunnel were linked.

CUTTING COAL

In an early underground coal mine, people had to dig the coal out with picks and shovels. Today, machines are used to cut coal from the coal face—the exposed coal on the walls of the mine (above).

DIGGING SMALL HOLES

This machine (above) is used to drill little holes into a coal mine wall, or coal face. The holes are loaded with explosives that are then set off. The wall shatters and falls to the floor. The rubble is gathered and removed. That's how we get coal!

TUNNEL BORING MACHINES

Creeping forward at a speed of 5 inches (12.7 centimeters) a minute, these long machines are used to dig tunnels. These tunnels can run under the seabed, through a mountain, or beneath busy city streets. At the front of a tunnel-boring machine (TBM), there is a huge cutter head that can rotate at different speeds to cut away the rock.

Push and pull

This motor makes sure that the cutter head is in the right position to cut the rock efficiently.

Cutter head

The cutter head is a rotating wheel that cuts rock. This cutting is less destructive than drilling and blasting, so tunnel boring machines can be used under big cities.

Drive motor

This huge electric motor spins the cutter head. The cutter head rotates about 1.5 to 3 times each minute.

Conveyor belt
This belt carries the rubble away from the tunnel face to waiting trucks.

Fixing the segments
The segments lock together snugly and the joints are filled. In some cases, the tunnel may need to be watertight, so segments are bolted together and rubber seals used.

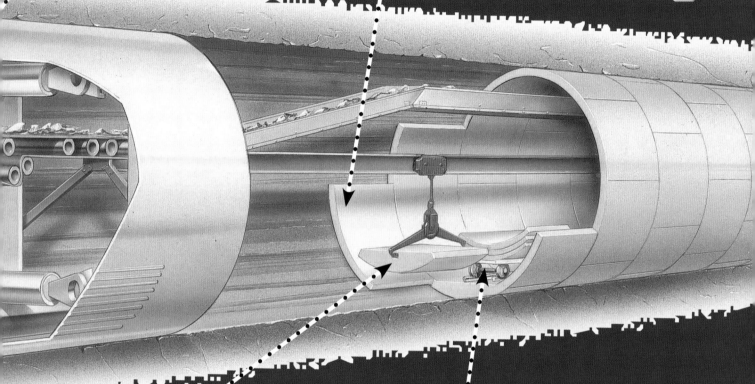

Concrete and iron segments
The tunnel is lined with concrete and iron segments, each weighing up to eight tons (7.26 metric tons)!

Carrying the segments
Each segment is lifted from its train car by a special crane. This crane then lifts the segments onto a trolley that carries them along the tunnel.

TUNNEL DIGGING

While tunnel-boring machines are quite expensive to construct and difficult to transport, for making long modern day tunnels, the cost and effort of using this massive machine is worth it for its efficiency.

Backhoe
The extendable hydraulic arm at the rear of a backhoe loader. It digs by using an inward movement.

Bucket
The scoop attachment on a digger that allows the machine to hold rubble.

Bulldozer
A strong machine with a blade on the front used for pushing materials where they need to go.

Caterpillar tracks
Wide belts that are fixed to a vehicle instead of wheels. They spread the weight of the vehicle over a large area and prevent it from sinking into soft ground.

Claw
A large grabbing attachment fixed on the end of a boom, often used for picking up large loads or tearing down a building.

Crane
A tall machine used for lifting heavy objects to high up places, usually with a projecting arm or beam.

Dredger
A machine made to scoop up dirt and sand from the bottom of a waterway in order to clear the passageway for water travel.

Engine
A machine with moving parts that converts power into motion that helps larger machines do their work.

Excavator
A large construction machine with a long boom, large scoop, and tracked wheels, often used for removing rubble from a construction site.

Foundation pile
A long, slender support placed in the ground of a tall building's foundation.

Grader
A machine with a long blade, used to create a flat surface for projects, such as road building or farming.

Hopper
A container that holds and helps distribute materials, such as the hopper on a road grader that holds asphalt just before it goes on the conveyor belt through the grader.

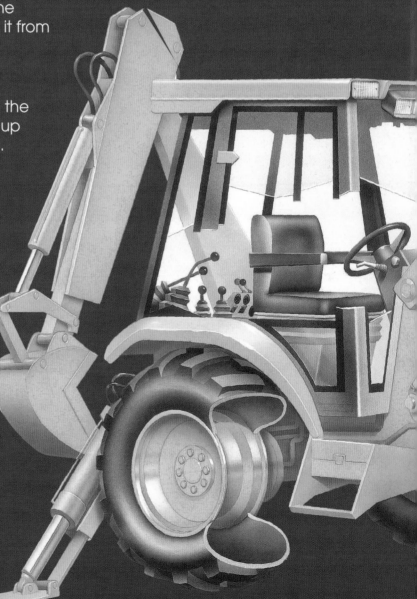

Hydraulic
This refers to objects that are moved or powered by a liquid, such as water or oil. Hydraulic rams are used to move the arms in most diggers.

Internal combustion engine
An engine where the burning of air and fuel occurs inside the engine's cylinders. Steam engines are not internal combustion engines because their fuel is burned outside the engine.

Mini excavator
An excavator with a small scoop and boom, made to work in tight conditions.

Mining shovel
A machine with a bucket used for digging and loading mined rock from the mine site.

Opencast mine
A mine close to the earth's surface, made by digging away layers of earth and rock, and removing valuable materials front the rock.

Piston
A rod that fits inside a cylinder and is moved up and down by the pressure of a gas or a liquid.

Pneumatic
This refers to objects that are moved or powered by compressed air, such as a pneumatic drill.

Sole plates
Plates on the back of a road paver that help smooth asphalt flat.

Stabilizers
Legs that keep a truck steady while it is working.

INDEX

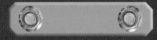
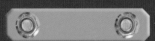

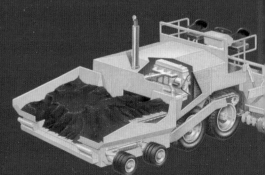

PHOTO CREDITS
Abbreviations: t-top, m-middle, b-bottom, r-right, l-left, c-center.
Pages 5 & 9 — Charles de Vere. 6 & 18t — Roger Vlitos. 6-7, 11t, 12l, & 13 all — Liebherr UK Ltd. 7t — JCB. 7b & 18b — Courtesy Finning UK Ltd. 10 all, 12b, 17, 19b, 23m, 26-27, & 27t — Frank Spooner Pictures. 14 both & 15 — Mary Evans Picture Library. 21 — Blaw Knox. 23t & 27m — Eye Ubiquitous. 27b — National Coal Board.